Maths Revision Booklet
for CCEA GCSE 2-tier specification

N6

Compiled by Joe McGurk

cea
Rewarding Learning

Contents

Revision Exercise 1a   (non-calculator)

You must **not** use a calculator for this paper. Total mark for this paper is 56.
Figures in brackets printed down the right-hand side of pages indicate the marks awarded to each question or part question.
You should have a ruler, compasses, set-square and protractor.

1

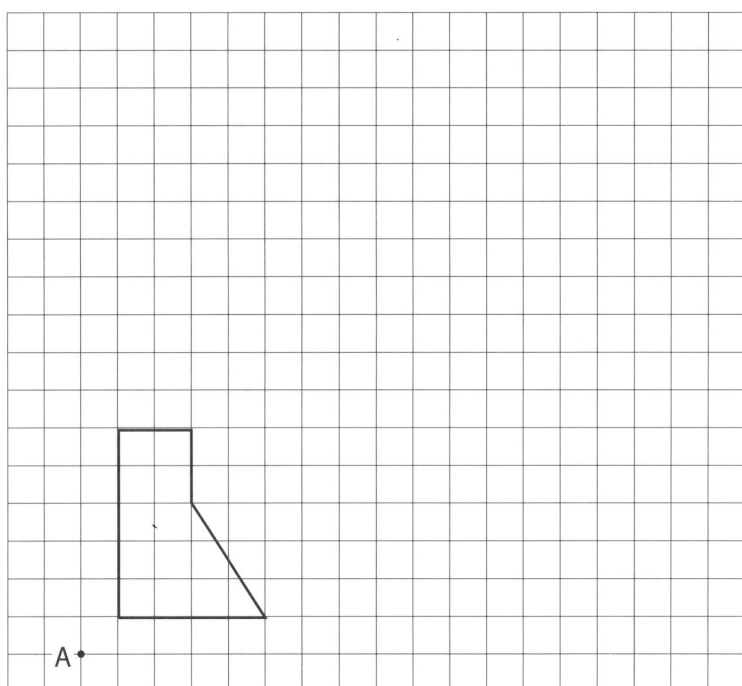

Enlarge the shape by a scale factor 2 about the point A.   [3]

2

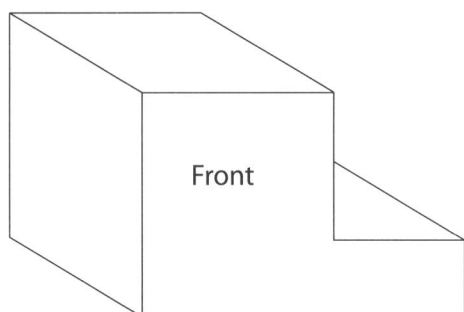

Draw the plan and the front elevation of the solid above.   [3]

Plan

Front Elevation

3 (a) Find the value of R when $S = -12$ and $T = -8$

$$R = \frac{6(S - T)}{2}$$

$R = \frac{6(-12+8)}{2}$ $R = \frac{-24}{2} = -12$

Answer $R = \underline{\quad -12 \quad}$ [3]

(b) Calculate the value of s in the formula below where $u = 40$, $a = -5$ and $t = 3$

$$s = ut + \frac{1}{2}at^2$$

$S = (40 \times 3) + \frac{1}{2}(-5 \times 3^2)$

$S = 120 + \cancel{38}^{-202.5} = -82.5$

Answer $s = \underline{\quad -82.5 \quad}$ [2]

4 Estimate the value of

$$\frac{509 \times 6}{(89.3 + 58.9)}$$

$\frac{510 \times 5}{90 + 60} = \frac{2,550}{150} = \frac{255}{15}$

Answer _____ [2]

5 A tin contains chocolate, nut, fruit and mint sweets. The probabilities of selecting each type of sweet at random from the tin are shown in the table below.

Type	Chocolate	Nut	Fruit	Mint
Probability	0.25	0.2	0.4	0.15

(a) What is the probability of selecting a mint sweet?

$p = 0.25 + 0.2 + 0.4 = 0.85$ $\therefore 1 - 0.85 = 0.15$

Answer $\underline{\quad 0.15 \quad}$ [2]

(b) What is the probability of **not** selecting a chocolate sweet?

$p(\text{not choc}) = 1 - \frac{1}{4} = \frac{3}{4}$

Answer $\underline{\quad \frac{3}{4} \quad}$ [2]

The tin contains 240 sweets.
(c) How many fruit sweets are there in the tin?

$0.1 = \frac{240}{10} = 24$

$\frac{0.4}{240}$

$= \frac{24 \times 4}{240} = \frac{96}{240} = \frac{48}{120} = \frac{24}{60} = \frac{12}{30} = \frac{6}{15}$

Answer $\underline{\quad \frac{6}{15} \quad}$ [2]

6 (a) Simplify $\dfrac{a^3 \times a^6}{a^2}$

$$\dfrac{a^9}{a^2} = a^7$$

Answer a^7 _____ [1]

(b) Calculate the reciprocal of 0.6

$$0.6 = \dfrac{3}{5} \qquad \therefore \dfrac{5}{3} = 1\dfrac{2}{3}$$

Answer $1\dfrac{2}{3}$ _____ [2]

7 A and B are two points which are 12 cm apart. By shading, find the locus of points which are less than 7 cm from A and such that the distance from A is greater than the distance from B.

[3]

8 l, r and h are lengths.

Write down whether each of the following formulae represents length, area, volume or none of these.

(a) $5\pi r^2 + \pi rl$

Answer _____*area*_____ [1]

area +

✳ **(b)** $6\pi r^2 h + \pi r\sqrt{(r^4 + h^4)}$

Answer _____*Volume*_____ [1]

9 (a) Complete the table below for $y = 2x^2 - 6x - 5$ for values of x from –2 to 5.

x	-2	-1	0	1	2	3	4	5
y	15	5	-5	-9	-9	-5	3	15

[2]

(b) Draw the graph of $y = 2x^2 - 6x - 5$ for values of x from $x = -2$ to $x = 5$

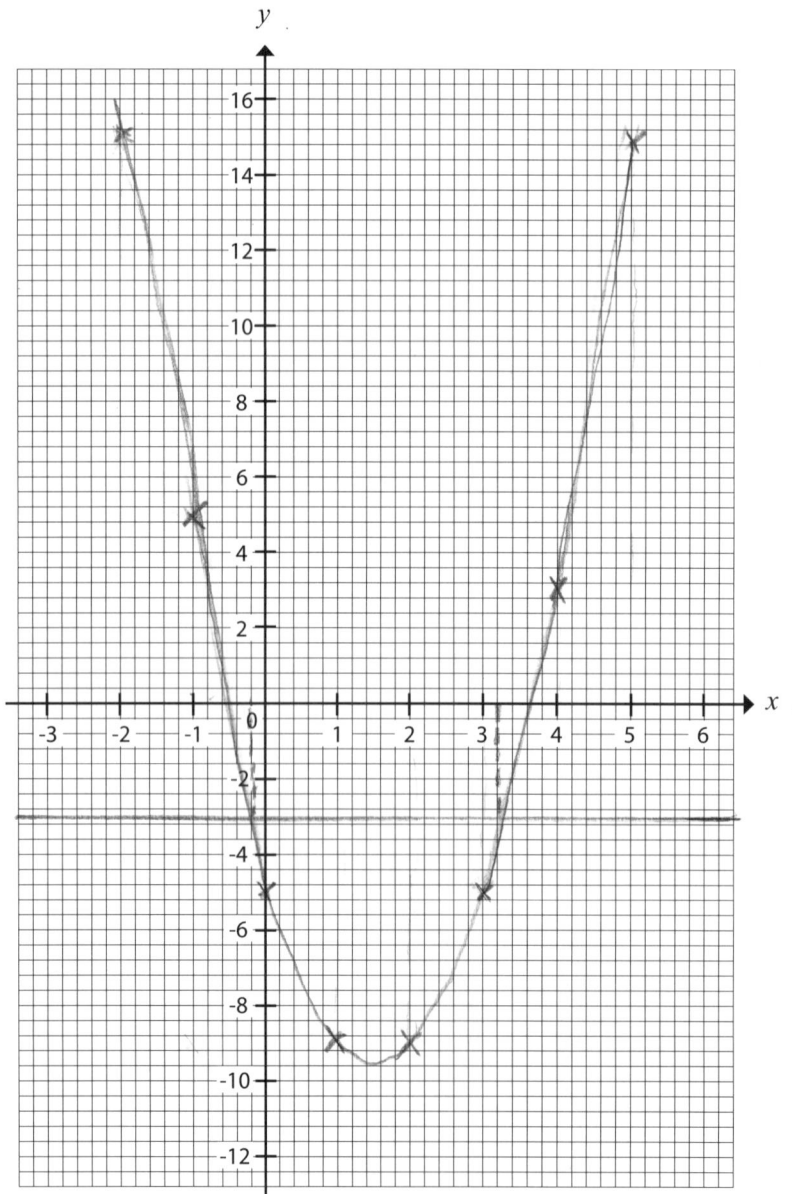

[2]

(c) Draw the line $y = -3$ and find the x values of the points of intersection of the line with $y = 2x^2 - 6x - 5$

Answer $x = \underline{-0.2}$ and $x = \underline{3.2}$ [2]

(d) Write down the equation in x with the solutions found in part **(c)** and simplify it.

$$y = mx + c$$

Answer _____ [1]

$$-3 = -0.2m + c$$

10 (a) Simplify $(\sqrt{5})^2$

$$\sqrt{5} \times \sqrt{5} = \sqrt{25}$$
$$= 2\sqrt{5}$$

Answer _____ [1]

(b) Express $\frac{4}{11}$ as a recurring decimal

Answer _____ [1]

(c) Rationalise the denominator of $\dfrac{5}{2\sqrt{3}}$

Answer _____ [2]

(d) Given $y^3 = \dfrac{20}{6x^2}$ and $y = 10$, find x in the form $a\sqrt{b}$

Answer _____ [3]

$$y^3 = \frac{20}{6x^2} \qquad 10^3 = \frac{20}{6x^2} \qquad 1000\,(6x^2) = 20$$
$$6x^2 = 20000$$
$$x^2 = 3333\tfrac{1}{3}$$

11 (a) Simplify the expression $(3x^2y^3z^4)^2$

$$\left(3x^2y^3z^4\right)\left(3x^2y^3z^4\right)$$

Answer _____ [2]

(b) Rearrange the expression to make c the subject

$$t = \frac{cy + s}{cx}$$

$$c = \frac{s}{t(x-y)}$$

$$t \times cx = cy + s$$

$$cx - cy = \frac{s}{t} \qquad c(x-y) = \frac{s}{t}$$

Answer _____ [3]

(c) simplify $(5 + 2\sqrt{10})^2$ $\qquad \left(5 + 2\sqrt{10}\right)\left(5 + 2\sqrt{10}\right)$

Answer _____ [2]

12

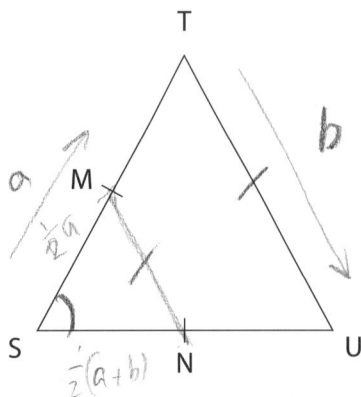

STU is an equilateral triangle. M and N are the midpoints of ST and SU respectively.

$\overrightarrow{ST} = \mathbf{a}$ and $\overrightarrow{TU} = \mathbf{b}$

Find in vector form:

(a) $\overrightarrow{SM}$

Answer $\frac{1}{2}\mathbf{a}$ _____ [1]

(b) $\overrightarrow{SU}$

Answer $\mathbf{a} + \mathbf{b}$ _____ [1]

(c) $\overrightarrow{SN}$

Answer $\frac{1}{2}(\mathbf{a} + \mathbf{b})$ _____ [1]

(d) Use vectors to show that $\overrightarrow{MN}$ is parallel to $\overrightarrow{TU}$ and half its length.

$\overrightarrow{MN} = \frac{1}{2}\mathbf{a} + \frac{1}{2}(\mathbf{a}+\mathbf{b})$

$\therefore \overrightarrow{MN} = \frac{1}{2}\mathbf{b}$

$T\hat{S}U$ is common angle.

[2]

13

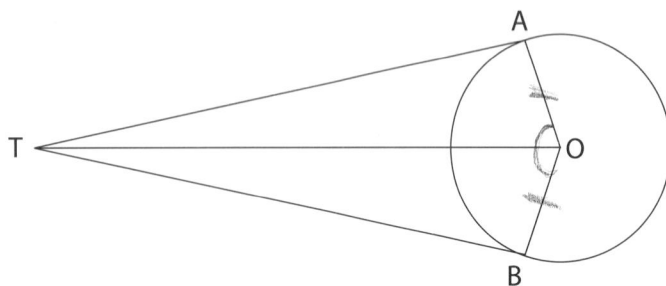

TA and TB are tangents to a circle, centre O.
By proving that triangles OAT and OBT are **congruent**, show that tangents TA and TB are equal.

$A\hat{O}B$ is a common angle

AO and BO are equal.

S SA

[3]

Revision Exercise 2a   (with calculator)

You may use a calculator for this paper. Total mark for this paper is 56.
Figures in brackets printed down the right-hand side of pages indicate the marks awarded to each question or part question.
You should have a ruler, compasses, set-square and protractor.

1 Which of "always even", "always odd", "could be odd or even" describes the number $n^2 + 1$ where n is an odd number. Explain.

Answer *"always even" because when you square an odd you get an odd, so by adding one you get an even.*   [2]

2 The recipe needed to make an energising sports drink is

Orange juice	100 ml
Sugar	56 g
Salt	12 g
Water	1900 ml

(a) John makes some of the sports drink. He uses 25 ml of orange juice. How much salt does he use?

$25ml = \frac{1}{4}$ $Salt = 12 \times \frac{1}{4}$
                          $= 3$

Answer _____3g_____ g [1]

(b) Sheila uses 84 g of sugar to make the sports drink. Find the amounts of orange juice, salt and water used.

orange juice _____ ml
salt _____ g
water _____ ml
[3]

3 Michael works in the Civil Service and earns an annual salary of £31000
The amount of tax which Michael pays is calculated from the taxable bands and the rates of tax shown in the table below.

Tax Bands	Tax rate
Personal allowance up to £5435	0%
£0 – £2230 taxable income	10%
£2231 – £34600 taxable income	22%

Calculate the annual amount of tax that Michael pays.

Answer £ _____ [3]

4

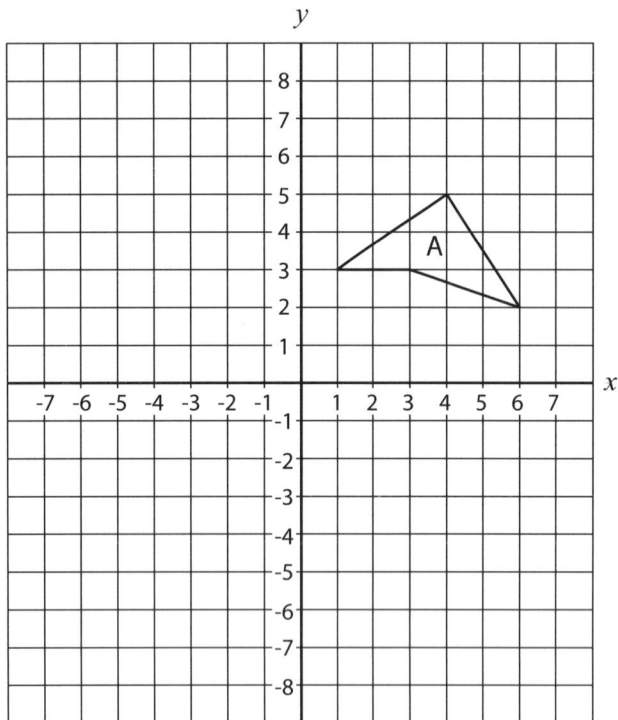

(a) Reflect shape A in the y axis. Label the reflection B.   [1]

(b) Reflect shape B in the x axis. Label this reflection C.   [1]

(c) Reflect B in the line $y = x$. Label this reflection D.   [2]

(d) Describe fully the single transformation which maps A onto C.

   Answer _____

   _____ [3]

5

Distance from
Belfast (miles)

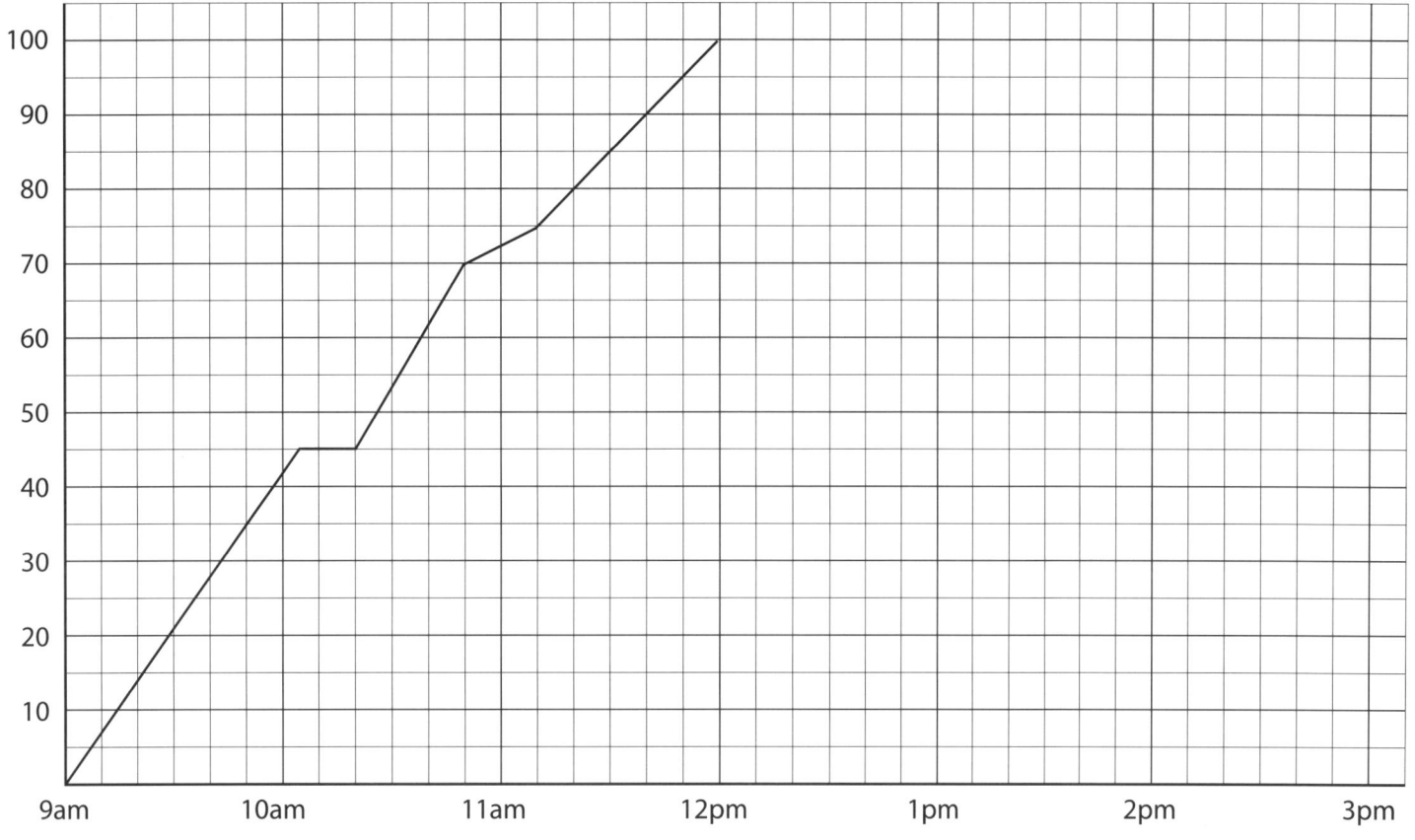

The graph represents Mr Brown's journey from Belfast to Donegal town. Mr Brown stopped for a break at a service station outside Dungannon.

(a) For how long did Mr Brown stop?

Answer _____ minutes [1]

Mr Brown resumed his journey but had to slow down outside Enniskillen because of heavy traffic.

(b) For how many miles did he travel at a slower average speed?

Answer _____ miles [1]

Mr Brown arrives in Donegal town at 12 noon. He attends a meeting for 40 minutes and returns non stop to Belfast at an average speed of 50 miles per hour.

(c) Complete the travel graph for Mr Brown's journey.   [2]

Mr Green leaves Belfast at 9.30am and drives directly to Donegal town at an average speed of 45 miles per hour.

(d) Draw the line showing Mr Green's journey.   [2]

(e) How far apart were Mr Brown and Mr Green at 9.50am?

Answer _____ miles [2]

(f) At what time did Mr Green overtake Mr Brown?

Answer _____ [2]

(g) How far were Mr Brown and Mr Green from Donegal town when Mr Green passed Mr Brown?

Answer _____ miles [1]

6 The density of beech wood is 740 kg/m^3. Calculate the mass of a block of beech which has a cross-section area of 0.19 m^2 and length 7.1 m. Give your answer to an **appropriate degree of accuracy**.

Answer _____ kg [4]

7 200 boys and 300 girls were asked to select their choice of fruit from an apple and an orange. The probability of a boy choosing an apple was $\frac{3}{5}$ and the probability of a girl choosing an apple was $\frac{3}{4}$
How many boys and girls chose an apple?

Answer _____ [4]

8 n is a number written in standard form.
When n is multiplied by 8.6×10^6, the result is 4.902×10^{11}
Find n.

Answer $n =$ _____ [2]

9 A hat contains 6 red balls and 4 blue balls.
Two balls are selected at random from the hat. The first ball is selected at random and is not replaced. A second ball is then selected from the hat.

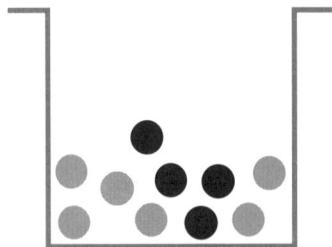

(a) Complete the probability tree diagram below to show the probabilities.

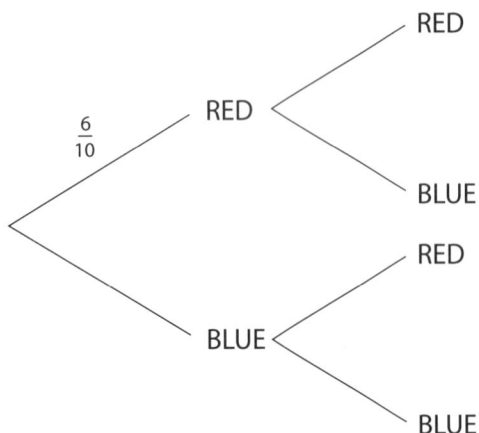

$\frac{6}{10}$

RED
    RED
    BLUE

BLUE
    RED
    BLUE

[2]

(b) What is the probability of selecting two blue balls?

Answer _____ [2]

(c) What is the probability of selecting at least one red ball?

Answer _____ [2]

10

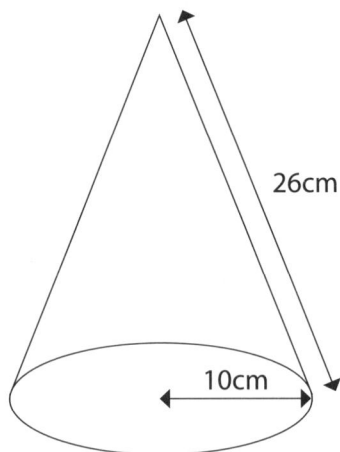

A right circular closed cone has a base radius of 10 cm and slant height 26 cm. Find the total surface area.

Answer _____ cm² [2]

11 The time (T) in weeks to complete a building job is inversely proportional to the square of the number of workers (w) who are employed on the job.
It takes 25 weeks for a particular job to be completed when 36 workers are employed.

(a) Find an equation connecting T and w.

Answer _____ [2]

(b) What is the minimum number of workers required so that the same job will be completed within 16 weeks?

Answer _____ [2]

12 A first year pharmacy student has to sit 3 examinations in 'Microbiology', 'Biochemistry' and 'Ethics and Practice'.

In Microbiology, the probability of her passing the examination is $\frac{8}{10}$

In Biochemistry, the probability of her passing the examination is $\frac{7}{10}$

In Ethics and Practice, the probability of her passing the examination is $\frac{9}{10}$

Assuming that the probability of the student passing any examination is independent of each of the other examinations, use a tree diagram or otherwise to calculate the probability that:

(a) (i) she will pass all three papers

                                                                        Answer _____ [2]

    (ii) she will fail all three papers

                                                                        Answer _____ [2]

    (iii) she will pass at least one of the papers

                                                                        Answer _____ [2]

The student needs to pass at least two of the three examinations to proceed into the second year of her studies.

    (b) Calculate the probability that she will succeed in going into the second year course.

                                                                        Answer _____ [2]

Revision Exercise **1b**   (non-calculator)

You must **not** use a calculator for this paper. Total mark for this paper is 56.
Figures in brackets printed down the right-hand side of pages indicate the marks awarded to each question or part question.
You should have a ruler, compasses, set-square and protractor.

1 (a) Describe fully the transformation which maps shape A onto shape B.

Answer _____

_____ [2]

(b) Enlarge shape A by a scale factor 2, centre (–4, –3)   [3]

(c) Rotate shape A by an angle of 90° clockwise about (0, –3)   [2]

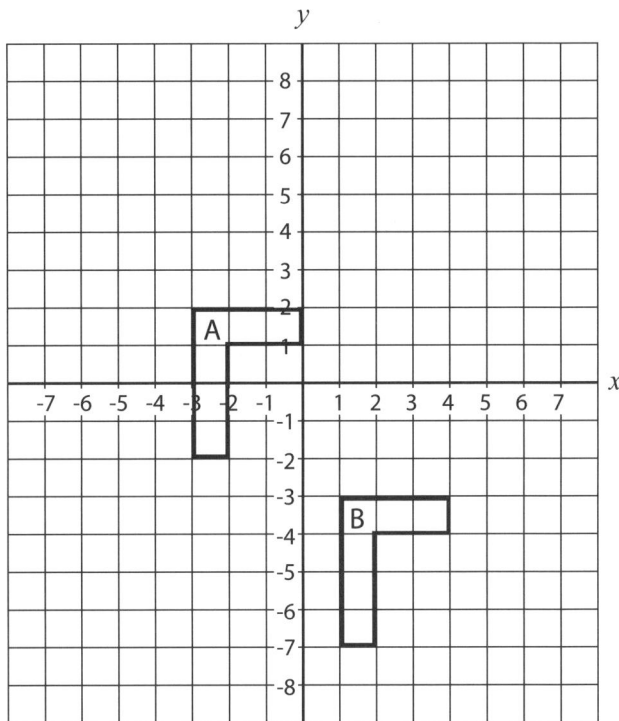

2 Find the value of M when $A = 12$ and $B = -8$

$$M = \frac{6(A - B)}{8}$$

Answer $M = $ _____ [3]

3 Estimate the value of

$$\frac{54.8 \times 4.03}{0.112}$$

Answer _____ [2]

4 Bags of sweets contain between 43 and 47 sweets in total. The probabilities of selecting bags containing each number of sweets are shown in the table below.

No. of sweets	43	44	45	46	47
Probability	0.05	0.15	0.5		0.1

(a) What is the probability of selecting a bag containing 46 sweets?

Answer _____ [2]

(b) What is the probability that a bag contains less than 45 sweets?

Answer _____ [2]

A box contains 500 bags of sweets.
(c) How many bags contain 44 sweets?

Answer _____ [2]

5 (a) Solve the inequality

$3x - 5 > -7$

Answer _____ [2]

(b) Make p the subject of the formula

$5r = 6m + p$

Answer _____ [2]

(c) Simplify
  (i) $\dfrac{t^2 \times t^5}{t^3}$

Answer _____ [1]

  (ii) $(2x^3y^4) \times (x^4y^2)$

Answer _____ [2]

6 (a) Explain why $2n - 1$ is always an odd number for any positive integer n.

Answer _____

_____ [2]

(b) Hence, by finding the square of $(2n - 1)$, prove that the square of an odd number is always odd.

[2]

7 A five sided spinner has the numbers 1, 2, 3, 4 and 5 labelled on each of its sides. The spinner is spun 500 times. The results are shown in the table below.

Number on spinner	1	2	3	4	5
Frequency	94	105	103	96	102

(a) What is the relative frequency of scoring 3?

Answer _____ [1]

(b) What is the relative frequency of scoring more than 2 on one spin of the spinner?

Answer _____ [2]

(c) Do the results in the table suggest that this is a fair spinner? Give a reason for your answer.

Answer _____

_____ [1]

8 Martha has a fair dice. She throws it twice.
What is the probability that she scores two sixes?

Answer _____ [2]

9 (a) Evaluate $(\sqrt{2})^4$

Answer _____ [1]

(b) Show that
$(\sqrt{8} + \sqrt{18})^2 = 50$

[2]

10 (a) Which of the following fractions is **not** a recurring decimal?

$\frac{1}{11}$ $\frac{3}{7}$ $\frac{5}{9}$ $\frac{7}{8}$ $\frac{5}{45}$

Answer _____ [1]

(b) Change the recurring decimal $0.\overset{..}{8}\overset{.}{1}$ into a fraction.

Answer _____ [2]

(c) Given that $p = \sqrt{8}$, $q = \sqrt{10}$ and $r = \sqrt{5}$, simplify $\frac{pq}{r}$

Answer _____ [2]

11 The letters r, l, d, h and w in all the expressions below represent lengths.
State whether each of the expressions represents a length, area, volume or none of these.

(a) $\pi r^2(r + l)$

Answer _____

(b) $\pi r l + 2d$

Answer _____

(c) $\dfrac{r \sqrt{(lw^3)}}{h}$

Answer _____

(d) $\dfrac{(hr^2 + dw^2)}{l^3}$

Answer _____

(e) $\dfrac{(\pi r l h + 4rw^2)}{rh}$

Answer _____ [3]

12

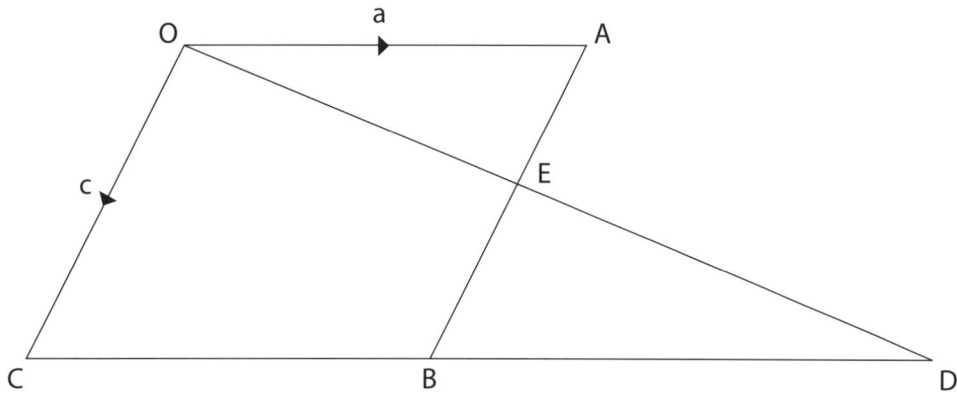

OABC is a parallelogram. The line CB is produced (extended) so that BC = BD

$\overrightarrow{OA}$ = **a**, $\overrightarrow{OC}$ = **c**

Find in terms of **a** and **c**:

(a) $\overrightarrow{CD}$

Answer _____ [1]

(b) $\overrightarrow{OD}$

Answer _____ [1]

(c) Given that E is the midpoint of AB, find the vector $\overrightarrow{OE}$

Answer _____ [1]

(d) Show, **using vectors**, that E is the midpoint of OD.

[2]

13

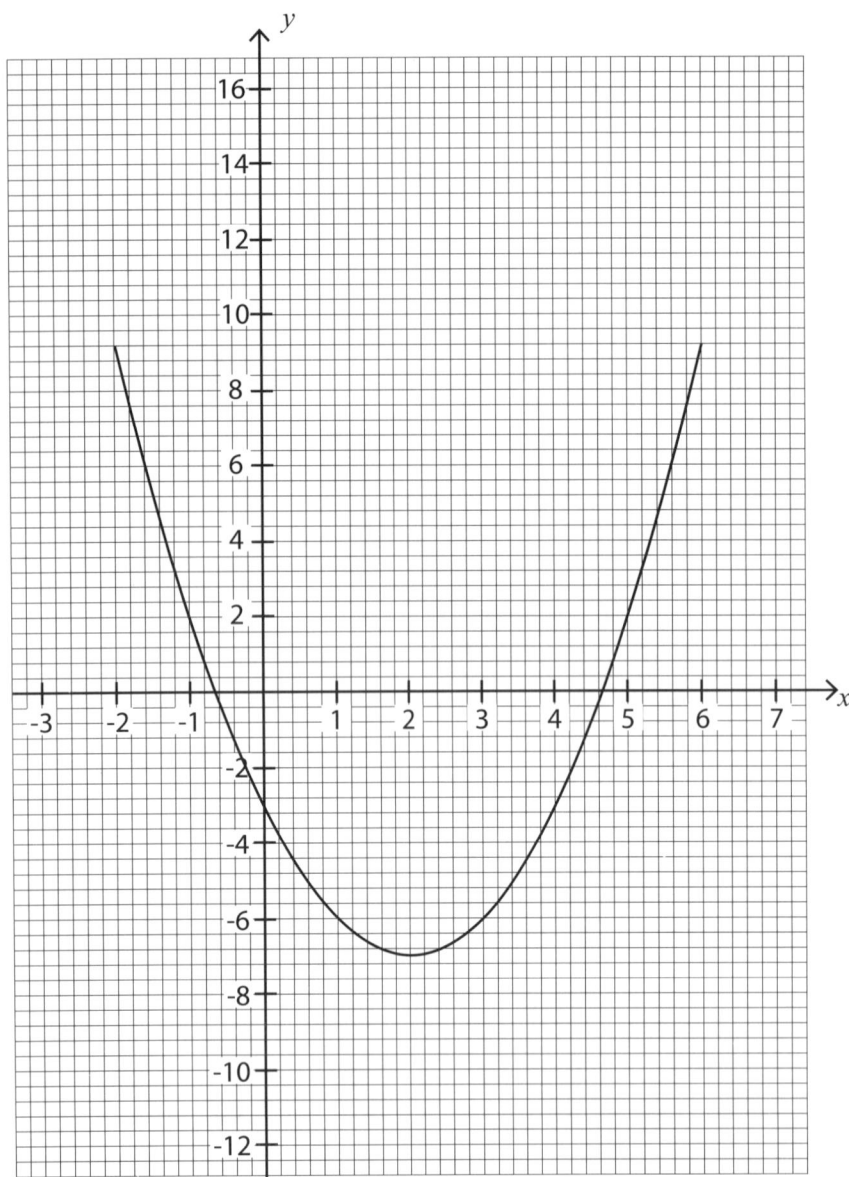

The graph of $y = x^2 - 4x - 3$ for values of x between $x = -2$ and $x = 6$ is drawn above.
By drawing an appropriate linear graph, find the solutions to the equation
$x^2 - 6x + 2 = 0$

Answer x = _____ *and x =* _____ [2]

14

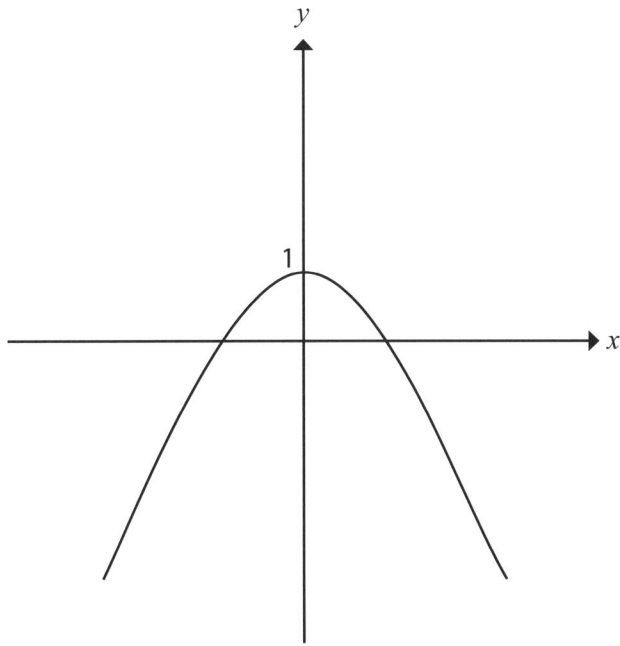

The diagram shows the graph of $y = f(x)$

(a) Sketch the graph of $y = f(x) - 2$

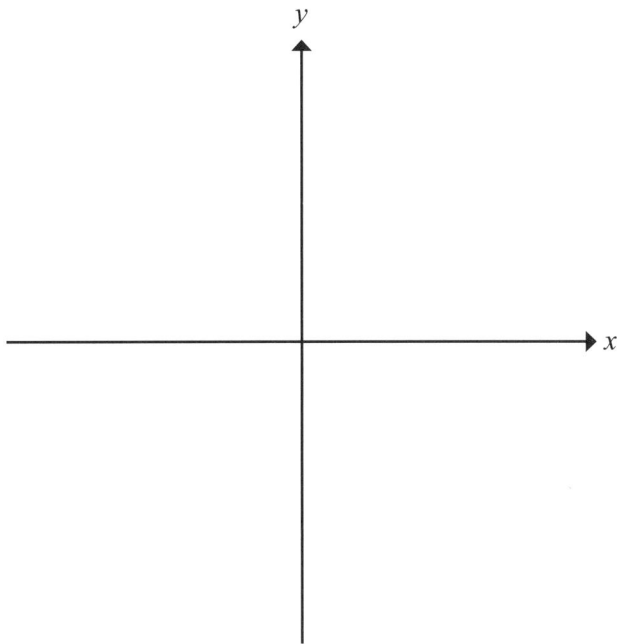

[1]

(b) Sketch the graph of $y = f(x - 3)$

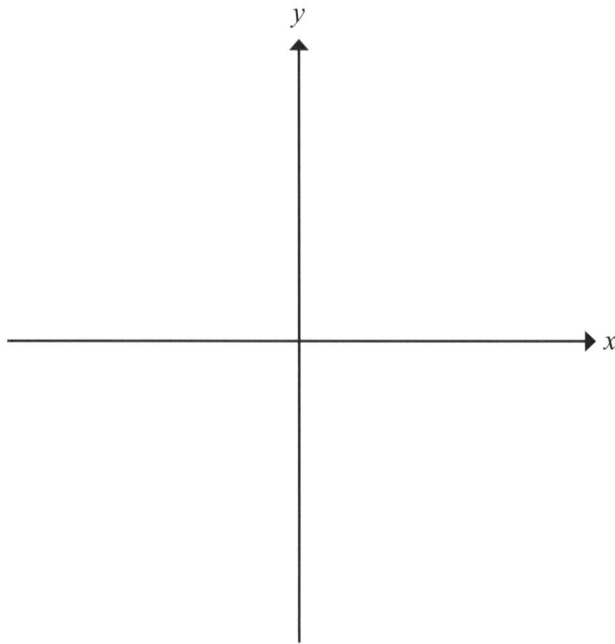

y

x

[1]

(c) Sketch the graph of $y = 3f(x)$

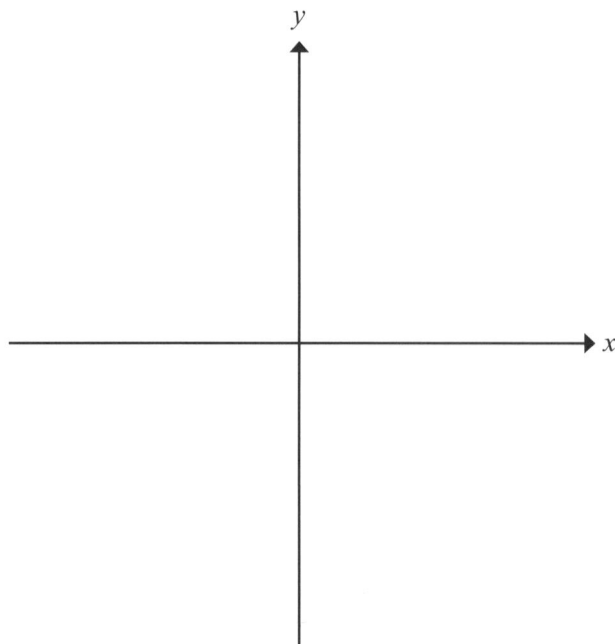

y

x

[1]

Revision Exercise 2b   (with calculator)

You may use a calculator for this paper. Total mark for this paper is 56.
Figures in brackets printed down the right-hand side of pages indicate the marks awarded to each question or part question.
You should have a ruler, compasses, set-square and protractor.

1

> ## SALE!
>
> 42" PLASMA SCREEN TELEVISION only £599
>
> or Hire Purchase terms: Pay £50 deposit and
> 24 monthly payments at £24.99

Janet decides to pay the sale price for the television as advertised.
How much does she save by not buying the television using Hire Purchase?

Answer £ _____ [2]

2

> ## SCHOOL UNIFORM
> Shirt £13.99
> Trousers £18.50
>
> SPECIAL OFFER!
> Buy both items and receive a 20% discount on the total price

Mark sees an advertisement in a shop window. Mark buys both items.
How much does he pay?

Answer £ _____ [3]

3 (a) Mark bought a computer game on his holidays in the United States. He paid $89.99
Susan bought the same game in Belfast for £34.99
The rate of exchange was £1 = $1.91
Who paid more for their computer game and by how much in pounds and pence?

Answer _____ by £ _____ [3]

(b) Divide £512 in the ratio 5:3

Answer _____ : _____ [2]

4 Jill bought a mobile phone and her price package was calculated as follows:
The cost of the line rental is £12.99 per month and calls are charged at 5.4 pence per minute.

(a) Write down a formula for C, the total cost in pounds for her monthly bill, where the length of time for calls is n minutes.

Answer _____ [2]

(b) Calculate Jill's total bill for a month when she had 637 minutes of calls.

Answer £ _____ [1]

(c) Jill's total bill for a particular month is £54.30. Calculate how many minutes of calls she made in this month.

Answer _____ minutes [2]

5

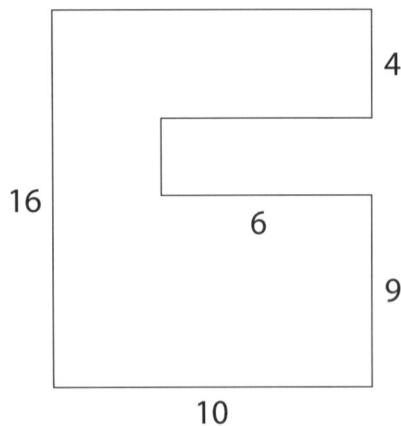

(a) Find the perimeter of the shape above where each of the dimensions are given in metres.

Answer Perimeter = _____ m [2]

(b) Find the area of the shape.

Answer Area = _____ m² [2]

(c)

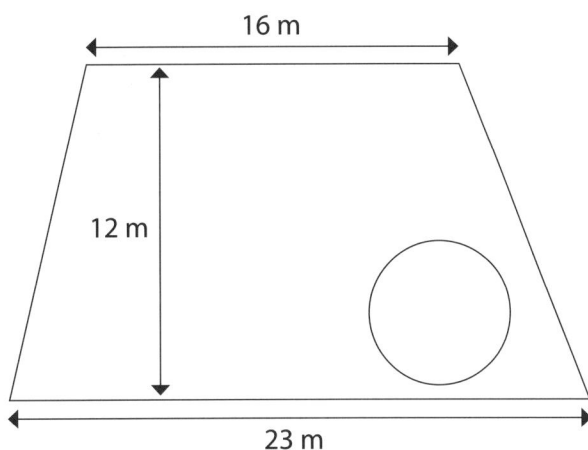

16 m

12 m

23 m

Tanya's back garden is in the shape of a trapezium.
The dimensions of the garden are shown in the diagram above.
The garden is paved apart from a circular pond of radius 3 m in one corner.
What area of the garden is paved?

Answer _____ m² [4]

[d]

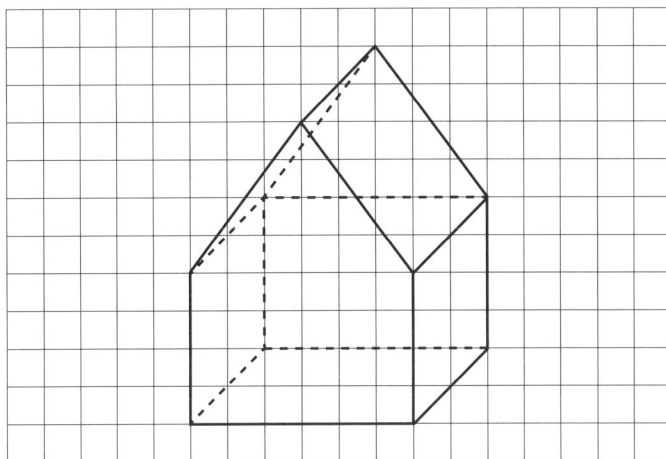

(i) Draw a plane of symmetry on the diagram above.   [1]

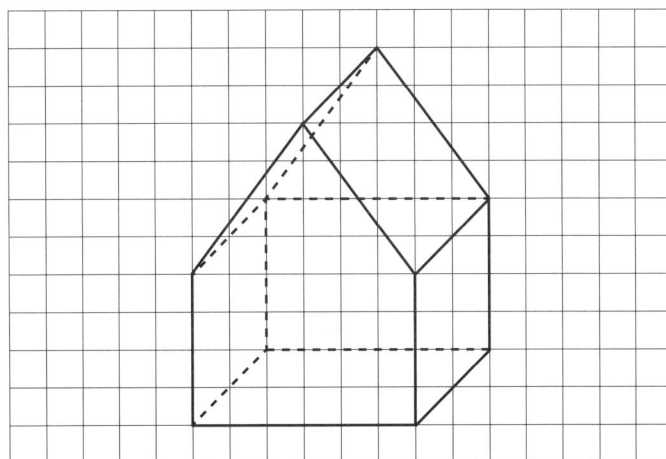

(ii) Draw a **different** plane of symmetry from part (i) on the diagram above.   [1]

6

Distance from
home (metres)

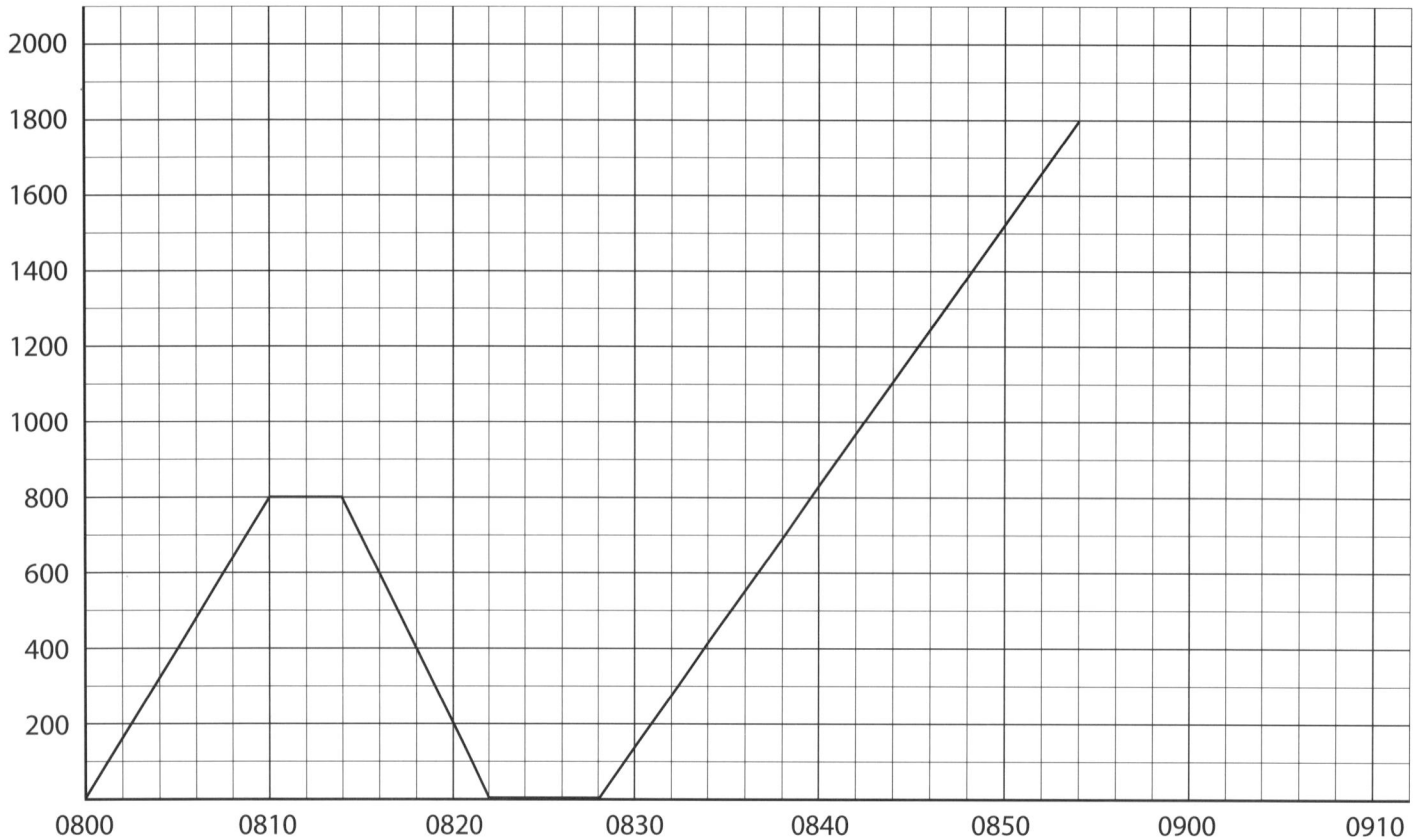

Sushi leaves home at 0800 to cycle to school. At 0810, she stops to buy sweets at the local shop and realises that she has left her money at home. She returns home to collect her money and eventually leaves home again at 0828 and cycles to school. The distance/time graph for her journey is shown above.

(a) How long was Sushi stopped at the shop?

Answer _____ minutes [1]

(b) What was Sushi's average speed in kilometres per hour between home and the shop?

Answer _____ km/hr [2]

Sushi's brother, Alan, leaves home at 0815 and walks to school at a constant speed of 2.4 km/hr.

(c) Draw the line which represents Alan's journey to school. [2]

(d) At what times do Alan and Sushi meet each other on the road?

Answer _____ and _____ [2]

(e) How far are Alan and Sushi from school when Sushi passes Alan?

Answer _____ metres [2]

7 The density of gold is 19300 kg/m^3. Calculate the mass, in grams, of 3 cm^3 of gold.

Answer _____ g [4]

8

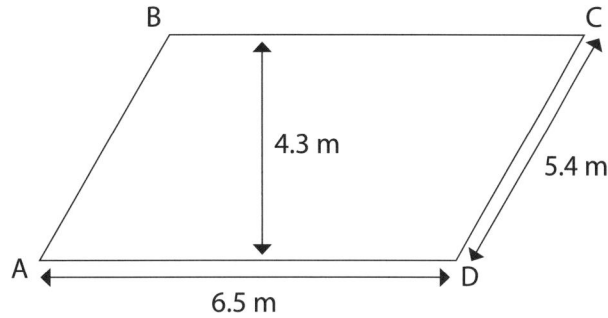

ABCD is a parallelogram with AD = 6.5 m and CD = 5.4 m. The perpendicular distance between the sides BC and AD is 4.3 m.
The parallelogram ABCD forms the base of the prism shown below.
The vertical height of the prism is 5.8 m.

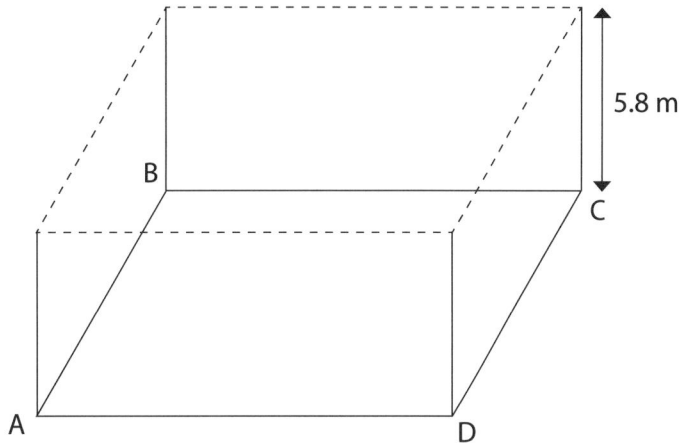

Find the volume of the prism.

Answer _____ [3]

9 Given that $L = \sqrt{(m/n)}$, find L in standard form when
$m = 8.1 \times 10^4$ and $n = 3.6 \times 10^8$

Answer _____ [3]

10 A cyclist experiences wind resistance which is directly proportional to the square of her speed. The resistance is 55 Newtons at a speed of 5 m/s.

(a) What will the resistance be at a speed of 15 m/s?

Answer _____ Newtons [2]

(b) At what speed is the cyclist travelling when the resistance is 198.55 Newtons?

Answer _____ m/s [1]

11

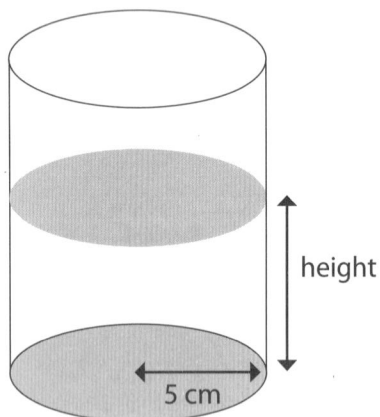

A cylindrical can of base radius 5 cm contains a soft drink. A volume of 125π cm³ of the drink is contained in the can. Find the surface area of liquid in contact with the can in terms of π .

Answer _____ cm² [3]

12 Kristin receives an iPod and a Playstation on his birthday.
The probability that the iPod will develop a fault in the first 3 years is 0.6
The probability that the Playstation will develop a fault in the first 3 years is 0.45
By drawing a tree diagram or otherwise, find the probability that
(a) neither device develops a fault within the first 3 years,

Answer _____ [3]

(b) only one of the devices develops a fault in the first 3 years.

Answer _____ [3]

ANSWERS

1a (non-calculator)

1

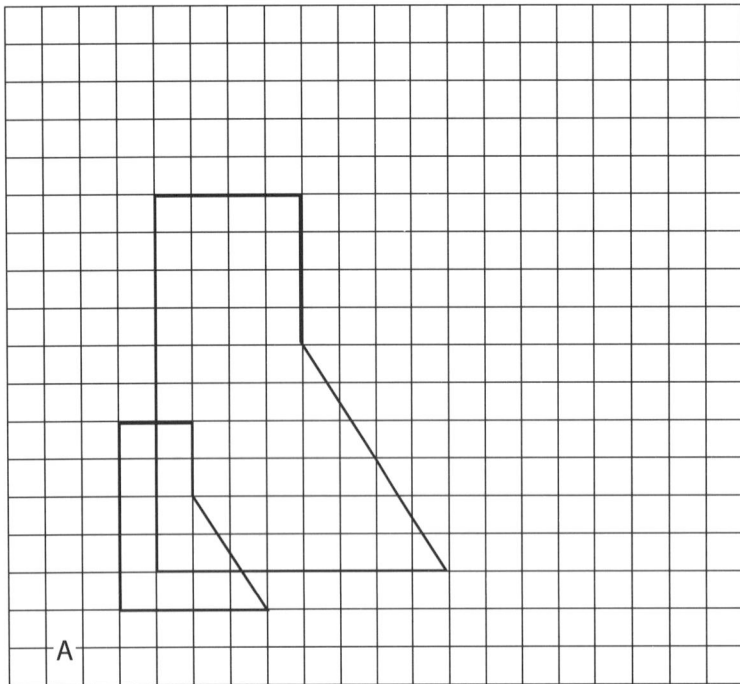

2

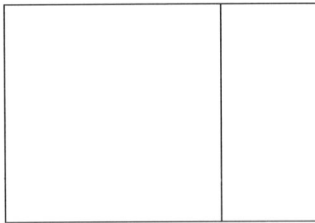

Plan

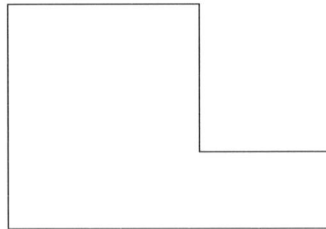

Front Elevation

3 (a) –12 **(b)** 97.5

4 20

5 (a) 0.15 **(b)** 0.75 **(c)** $0.4 \times 240 = 96$

6 (a) a^7 **(b)** $\frac{10}{6}$ or $\frac{5}{3}$ or 1.66 …

7

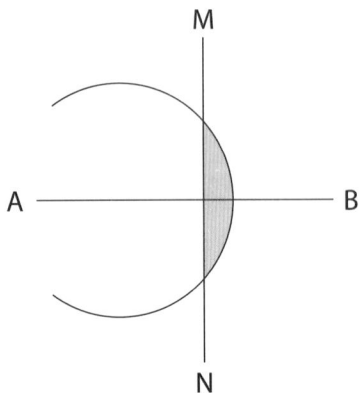

All points less than 7 cm from A are inside the circle, centre A, radius 7 cm. MN is the perpendicular bisector of AB. Hence, the locus is the shaded area as shown.

8 (a) area **(b)** volume

9 (a) 3, –9 **(b)** points plotted, smooth curve (not straight line segments)
(c) –0.3, 3.3 approx **(d)** $2x^2 - 6x - 5 = -3 \longrightarrow 2x^2 - 6x - 2 = 0$

10 (a) 5 **(b)** 0.363636…. **(c)** $\frac{5\sqrt{3}}{6}$ **(d)** $\frac{\sqrt{3}}{30}$

11 (a) $9x^4y^6z^8$ **(b)** $cxt - cy = s \longrightarrow c(xt - y) = s \longrightarrow c = \dfrac{s}{(xt - y)}$

(c) $25 + 20\sqrt{10} + 40 = 65 + 20\sqrt{10}$

12 $\frac{1}{2}\mathbf{a}$, $\mathbf{a} + \mathbf{b}$, $\frac{1}{2}(\mathbf{a} + \mathbf{b})$, $\overrightarrow{MN} = \overrightarrow{MS} + \overrightarrow{SN} = -\frac{1}{2}\mathbf{a} + \frac{1}{2}(\mathbf{a} + \mathbf{b}) = \frac{1}{2}\mathbf{b}$

$\overrightarrow{MN}$ is parallel to $\overrightarrow{TU}$ as $\overrightarrow{MN} = k\,\overrightarrow{TU}$ so $\overrightarrow{MN} = \frac{1}{2}\mathbf{b} = \frac{1}{2}\overrightarrow{TU}$

13 <OAT = <OBT = 90° (tangent to radius right angle)
OA = OB (radii of circle)
OT is the hypotenuse of both triangles OAT and OBT
Therefore, the triangles are congruent (using RHS property)
TA = TB

2a (with calculator)

1 Always even, since n^2 is odd as odd × odd = odd

2 (a) 3 **(b)** 150, 18, 2850

3 31000 – 5435 = 25565
10% of £2230 = £223
25565 – 2230 = £23335
22% of £23335 = £5133.70
Total tax paid = 5133.70 + 223 = £5356.70

4

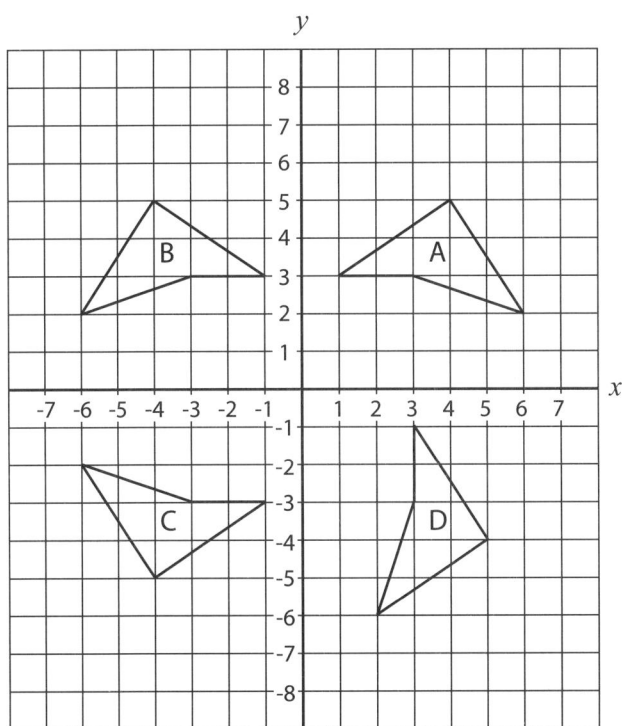

(d) C rotates A through 180° about the origin.

5 (a) 15 min **(b)** 5 **(c)** horizontal line from 12 to 12.40 at 100 miles, line from (12.40, 100) to 2.40pm on horizontal axis **(d)** line from 9.30 through (10.30, 45) and (11.30, 90) **(e)** Readings at 35 and 15 miles giving 20 miles **(f)** 11.10am **(g)** 25

6 998 = 1000 (appropriate degree of accuracy)

7 $\frac{3}{5} \times 200 + \frac{3}{4} \times 300 = 120 + 225 = 345$

8 $(4.902 \times 10^{11}) \div (8.6 \times 10^6) = 5.7 \times 10^4$

9 (a)

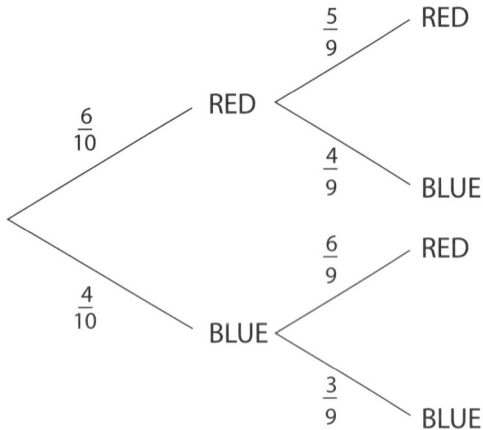

(b) $\frac{4}{10} \times \frac{3}{9} = \frac{12}{90} = \frac{2}{15}$

(c) $1 - \frac{2}{15} = \frac{13}{15}$

10 $\pi r l + \pi r^2 = (360\pi)\ 1131\,\text{cm}^2$

11 (a) $T = k/w^2$ giving $k = 32400$ **(b)** 45 workers

12 (a) (i) 0.504 **(ii)** 0.006 **(iii)** 1 − 0.006 = 0.994 **(b)** 0.902

1b (non-calculator)

1 (a) Translation $\begin{pmatrix} 4 \\ -5 \end{pmatrix}$

(b) and **(c)**

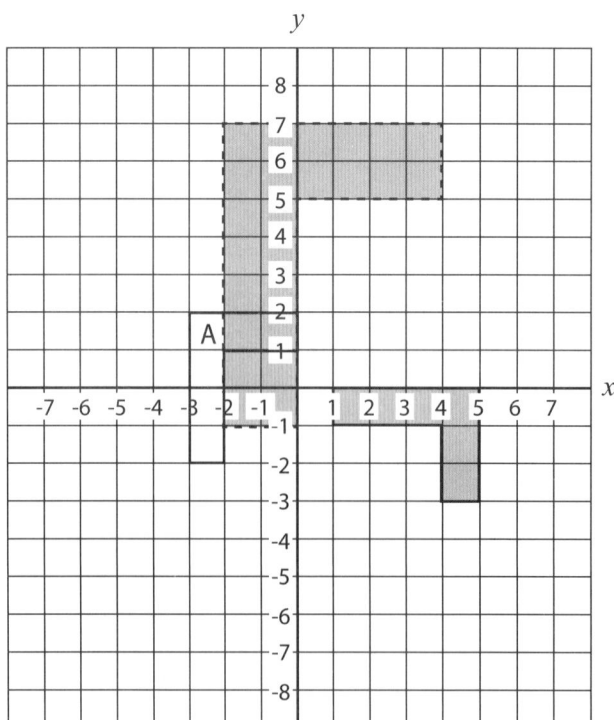

2 15

3 2000

4 (a) 0.2 **(b)** 0.2 **(c)** $0.15 \times 500 = 75$

5 (a) $x > -\frac{2}{3}$ **(b)** $p = 5r - 6m$ **(c) (i)** t^4 **(ii)** $2x^7y^6$

6 (a) Two times n is always an even number for any positive integer n, so by subtracting 1 we always get an odd number

(b) $(2n - 1)^2 = 4n^2 - 4n + 1$

   $4n^2$ is even as 4 multiplied by any integer is even.

   Similarly $4n$ is even.

   $4n^2 - 4n$ is even as an even number minus an even number is even.

   Hence, $4n^2 - 4n + 1$ is odd (by adding 1 to an even number)

7 (a) $\frac{103}{500}$ **(b)** $\frac{301}{500}$ **(c)** Yes, as relative frequencies are nearly all 0.2

8 $\frac{1}{36}$

9 (a) 4 **(b)** $8 + 2\sqrt{144} + 18 = 50$

10 (a) $\frac{7}{8}$ **(b)** $\frac{9}{11}$ **(c)** 4

11 (a) volume **(b)** none **(c)** area **(d)** none **(e)** length

12 (a) $2\mathbf{a}$ **(b)** $2\mathbf{a} + \mathbf{c}$ **(c)** $\overrightarrow{OE} = \mathbf{a} + \frac{1}{2}\mathbf{c}$ **(d)** $\overrightarrow{OE} = \mathbf{a} + \frac{1}{2}\mathbf{c} = \frac{1}{2}(2\mathbf{a} + \mathbf{c}) = \frac{1}{2}\overrightarrow{OD}$

13 $x^2 - 6x + 2 = x^2 - 4x - 3 - 2x + 5$

Hence, $x^2 - 6x + 2 = 0$ can be written
$$x^2 - 4x - 3 - 2x + 5 = 0 \quad \text{or}$$
$$x^2 - 4x - 3 = 2x - 5$$
Graph of $y = x^2 - 4x - 3$ is shown on diagram.
Draw the graph of $y = 2x - 5$ and find the points of intersection.
$x = 0.35$ and $x = 5.6$ approximately.

14 (a) shifted down 2 units crossing y axis at −1 **(b)** shifted right 3 units **(c)** stretched along the y axis scale factor 3, cuts at (0, 3) (graph slimmer)

2b (with calculator)

1 $24 \times 24.99 = 599.76 + 50 = 649.76$

   Saving $= 649.76 - 599 = £50.76$

2 £26.00 (25.99)

3 (a) $89.99 \div 1.91 = £47.11$

   $47.11 - 34.99 = 12.12$

   Mark paid £12.12 more for his game.

(b) £320 : £192

4 (a) $C = 12.99 + 0.054n$ **(b)** £47.39 **(c)** 765 minutes

5 (a) Perimeter $= 64$ **(b)** Area $= 142$

(c) Area of trapezium $= \frac{1}{2}(16 + 23) \times 12 = 234 \text{ m}^2$

   Area of pond $= \pi r^2 = \pi (3)^2 = 28.3 \text{ m}^2$

   Area of paving $= 234 - 28.3 = 205.7 \text{ m}^2$

(d) (i)

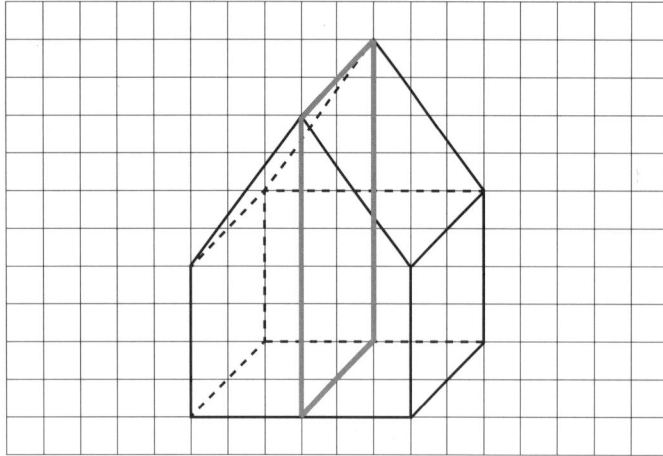

(ii)

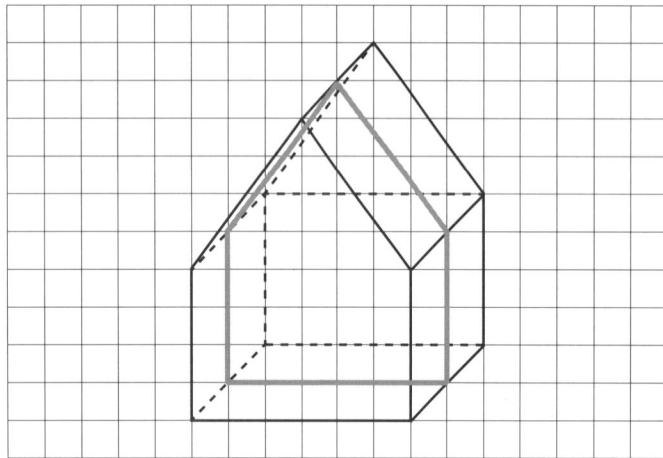

6 (a) 4 min **(b)** 800 m in 10 min = 4800 m in 60 min = 4.8 km/hr **(c)** line from 0815 finishing at (0900, 1800) **(d)** 0820 and 0846 **(e)** 1800 – 1250 (approx) = 550m

7 57.9 grams

8 V = 6.5 x 4.3 x 5.8 = 162.11 m³ (mark awarded for units m³ here)

9 $L = \sqrt{(2.25 \times 10^{-4})} = 1.5 \times 10^{-2}$

10 (a) $R = kv^2$ giving $k = 2.2$ so $R = 2.2(15)^2 = 495$ **(b)** 9.5

11 $\pi r^2 h = 125\pi$ gives $h = 5$cm

    S.A = πr^2 (base) + $2\pi rh$ (curved surface) = 75π

12 (a) 0.22 **(b)** 0.33 + 0.18 = 0.51